ENERGY SECTOR STANDARD
OF THE PEOPLE'S REPUBLIC OF CHINA

中华人民共和国能源行业标准

Technical Code for Prototype Observation
of Water Temperature for Hydropower Projects

水电工程水温原型观测技术规范

NB/T 10142-2019

Chief Development Department: China Renewable Energy Engineering Institute

Approval Department: National Energy Administration of the People's Republic of China

Implementation Date: October 1, 2019

China Water & Power Press

中国水利水电出版社

Beijing 2024

All rights reserved. No part of this publication may be reproduced, stored in a retrieval system, or transmitted in any form or by any means—electronic, mechanical, photocopying, recording or otherwise, without prior written permission of the publisher.

图书在版编目（CIP）数据

水电工程水温原型观测技术规范：NB/T 10142-2019 = Technical Code for Prototype Observation of Water Temperature for Hydropower Projects (NB/T 10142-2019)：英文 / 国家能源局发布. -- 北京：中国水利水电出版社, 2024. 8. -- ISBN 978-7-5226-2709-0

Ⅰ. TV72-65

中国国家版本馆CIP数据核字第2024E7Q098号

ENERGY SECTOR STANDARD
OF THE PEOPLE'S REPUBLIC OF CHINA
中华人民共和国能源行业标准

Technical Code for Prototype Observation
of Water Temperature for Hydropower Projects
水电工程水温原型观测技术规范
NB/T 10142-2019
（英文版）

Issued by National Energy Administration of the People's Republic of China
国家能源局　发布
Translation organized by China Renewable Energy Engineering Institute
水电水利规划设计总院　组织翻译
Published by China Water & Power Press
中国水利水电出版社　出版发行
　　Tel: (+ 86 10) 68545888　68545874
　　sales@mwr.gov.cn
　　Account name: China Water & Power Press
　　Address: No.1, Yuyuantan Nanlu, Haidian District, Beijing 100038, China
　　http: //www.waterpub.com.cn
中国水利水电出版社微机排版中心　排版
北京中献拓方科技发展有限公司　印刷
184mm×260mm　16开本　2印张　63千字
2024年8月第1版　2024年8月第1次印刷
Price(定价)：￥455.00

Introduction

This English version is one of China's energy sector standard series in English. Its translation was organized by China Renewable Energy Engineering Institute authorized by National Energy Administration of the People's Republic of China in compliance with relevant procedures and stipulations. This English version was issued by National Energy Administration of the People's Republic of China in Announcement [2023] No. 4 dated May 26, 2023.

This version was translated from the Chinese standard NB/T 10142-2019, *Technical Code for Prototype Observation of Water Temperature for Hydropower Projects*, published by China Water & Power Press. The copyright is reserved by National Energy Administration of the People's Republic of China. In the event of any discrepancy in the implementation, the Chinese version shall prevail.

Many thanks go to the staff from the relevant standard development organizations and those who have provided generous assistance in the translation and review process.

For further improvement of the English version, any comments and suggestions are welcome and should be addressed to:

China Renewable Energy Engineering Institute
No. 2 Beixiaojie, Liupukang, Xicheng District, Beijing 100120, China
Website: www.creei.cn

Translating organizations:

China Renewable Energy Engineering Institute

POWERCHINA Guiyang Engineering Corporation Limited

Translating staff:

CHANG Li	YU Weiqi	ZHANG Junwei	GAO Guoqing
ZHANG Shuyu	ZHOU Chao	LU Bo	ZHAO Zaixing
CHEN Fan	YI Zhongqiang	ZHANG Qian	

Review panel members:

LIU Xiaofen	POWERCHINA Zhongnan Engineering Corporation Limited
JIN Feng	Tsinghua University

QIAO Peng	POWERCHINA Northwest Engineering Corporation Limited
QIE Chunsheng	Senior English Translator
ZHANG Ming	Tsinghua University
YAN Wenjun	Army Academy of Armored Forces, PLA
JIA Haibo	POWERCHINA Kunming Engineering Corporation Limited
GAO Yan	POWERCHINA Beijing Engineering Corporation Limited
LI Shisheng	China Renewable Energy Engineering Institute

National Energy Administration of the People's Republic of China

翻译出版说明

本译本为国家能源局委托水电水利规划设计总院按照有关程序和规定，统一组织翻译的能源行业标准英文版系列译本之一。2023 年 5 月 26 日，国家能源局以 2023 年第 4 号公告予以公布。

本译本是根据中国水利水电出版社出版的《水电工程水温原型观测技术规范》NB/T 10142—2019 翻译的，著作权归国家能源局所有。在使用过程中，如出现异议，以中文版为准。

本译本在翻译和审核过程中，本标准编制单位及编制组有关成员给予了积极协助。

为不断提高本译本的质量，欢迎使用者提出意见和建议，并反馈给水电水利规划设计总院。

  地址：北京市西城区六铺炕北小街 2 号
  邮编：100120
  网址：www.creei.cn

本译本翻译单位：水电水利规划设计总院
       中国电建集团贵阳勘测设计研究院有限公司

本译本翻译人员：常　理　喻卫奇　张峻玮　高国庆
       张蜀豫　周　超　陆　波　赵再兴
       陈　凡　易仲强　张　倩

本译本审核人员：

  刘小芬　中国电建集团中南勘测设计研究院有限公司
  金　峰　清华大学
  乔　鹏　中国电建集团西北勘测设计研究院有限公司
  郗春生　英语高级翻译
  张　明　清华大学
  闫文军　中国人民解放军陆军装甲兵学院
  贾海波　中国电建集团昆明勘测设计研究院有限公司
  高　燕　中国电建集团北京勘测设计研究院有限公司
  李仕胜　水电水利规划设计总院

国家能源局

Announcement of National Energy Administration of the People's Republic of China
[2019] No. 4

National Energy Administration of the People's Republic of China has approved and issued 297 sector standards such as *Code for Electrical Design of Photovoltaic Power Projects*, including 105 energy standards (NB), 168 electric power standards (DL), and 24 petrochemical standards (NB/SH).

Attachment: Directory of Sector Standards

National Energy Administration of the People's Republic of China

June 4, 2019

Attachment:

Directory of Sector Standards

Serial number	Standard No.	Title	Replaced standard No.	Adopted international standard No.	Approval date	Implementation date
...						
15	NB/T 10142-2019	Technical Code for Prototype Observation of Water Temperature for Hydropower Projectss			2019-06-04	2019-10-01
...						

Foreword

According to the requirements of Document GNKJ [2015] No. 283 issued by National Energy Administration of the People's Republic of China, "Notice on Releasing the Development and Revision Plan of the Energy Sector Standards in 2015", and after extensive investigation and research, summarization of practical experience, consultation of relevant advanced standards, and wide solicitation of opinions, the drafting group has prepared this code.

The main technical contents of this code include: general provisions, basic requirements, basic data, technical requirements for observation, observation instruments and equipment, and observation data.

National Energy Administration of the People's Republic of China is in charge of the administration of this code. China Renewable Energy Engineering Institute has proposed this code and is responsible for its routine management. Energy Sector Standardization Technical Committee on Hydropower Planning, Resettlement and Environmental Protection is responsible for the explanation of specific technical contents. Comments and suggestions in the implementation of this code should be addressed to:

China Renewable Energy Engineering Institute
No. 2 Beixiaojie, Liupukang, Xicheng District, Beijing 100120, China

Chief development organizations:

China Renewable Energy Engineering Institute

POWERCHINA Guiyang Engineering Cooperation Limited

Participating development organizations:

POWERCHINA Northwest Engineering Cooperation Limited

Longtan Hydropower Development Limited Liability Company

Sichuan University

Chief drafting staff:

CHANG Li	LU Bo	WEI Lang	WANG Haiwen
YU Weiqi	ZHAO Zaixing	NIU Le	QI Yanbin
AN Ruidong	TANG Zhongbo	JIANG Hao	TUO Youcai
TANG Xing	CHEN Fan	LIU Rui	ZONG Xiao
YI Zhongqiang	ZHOU Chao	LI Nan	WANG Huoyun

ZHANG Junwei	ZHANG Qian	FAN Xinke	WU Yi
WANG Xiangyu	ZHAO Zhangguo	GUO Quan	DING Hongsheng
GAO Guoqing			

Review panel members:

WAN Wengong	XUE Lianfang	CHEN Guozhu	RUI Jianliang
CHEN Yuying	CHEN Min	CHEN Bangfu	LI Jinghua
NIU Tianxiang	JIN Yi	DAI Xiangrong	WEI Fan
QIANG Jihong	HUANG Bin	LEI Shaoping	CHEN Wenhua
LI Shisheng			

Contents

1	**General Provisions**	**1**
2	**Basic Requirements**	**2**
3	**Basic Data**	**3**
4	**Technical Requirements for Observation**	**4**
4.1	General Requirements	4
4.2	Water Temperature Observation of Reservoir Inflow	4
4.3	Vertical Water Temperature Observation of Reservoir	4
4.4	Water Temperature Observation for Selective Water Withdrawal Effectiveness	6
4.5	Water Temperature Observation of Reservoir Outflow and Downstream Reaches	7
5	**Observation Instruments and Equipment**	**9**
5.1	General Requirements	9
5.2	Configuration of Instruments and Equipment	9
5.3	Management of Instruments and Equipment	10
6	**Observation Data**	**11**
6.1	Observation Data Recording	11
6.2	Observation Data Compilation	11
6.3	Result Analysis and Data Management	11
Appendix A	**Contents of Report on Prototype Observation of Water Temperature for Hydropower Projects**	**13**
Appendix B	**Record Form for Prototype Observation of Water Temperature for Hydropower Projects**	**15**
Explanation of Wording in This Code		**16**
List of Quoted Standards		**17**

1 General Provisions

1.0.1 This code is formulated with a view to standardizing the scope and methods and unifying the technical requirements for the prototype observation of water temperature for hydropower projects.

1.0.2 This code is applicable to the prototype observation of water temperature for hydropower projects.

1.0.3 The prototype observation of water temperature for hydropower projects shall be based on the reservoir type and ecological protection requirements, follow the principles of systematicness, coordination and representativeness, formulate an observation scheme, and achieve technical feasibility, economic rationality, safety and reliability. The observation results shall reflect the distribution characteristics and variation pattern of water temperature of hydropower project.

1.0.4 In addition to this code, the prototype observation of water temperature for hydropower projects shall comply with other current relevant standards of China.

2 Basic Requirements

2.0.1 The prototype observation of water temperature for a hydropower project shall fully consider the characteristics of the reservoir and the way they impact water temperature, and shall pay attention to the requirements for water temperature of different purposes and environmentally sensitive objects.

2.0.2 For the prototype observation of water temperature for a hydropower project, technical requirements for the observation condition, range, period, frequency, section, line, point and method shall be reasonably determined according to the reservoir type and observation tasks, and field observation shall be carried out as per the scheme. If initial observation results fail to reflect the water temperature distribution characteristics, the observation scheme shall be adjusted timely. In the case of selective water withdrawal management, a hydrodynamic observation shall be conducted simultaneously.

2.0.3 Prototype observation of water temperature for a hydropower project shall rely on the existing water temperature observation stations, hydrologic stations, stage gauging stations, water quality monitoring stations, and waterward structures, and take into account such factors as accessibility, communications, maintenance, and safety.

2.0.4 For prototype observation of water temperature for a hydropower project, the observation data shall be processed and compiled, and a prototype observation report shall be prepared. The contents of report on prototype observation of water temperature for hydropower projects shall comply with Appendix A of this code.

3 Basic Data

3.0.1 For the prototype observation of water temperature for a hydropower project, basic data shall be collected sufficiently to meet the requirements of the observation tasks.

3.0.2 For the prototype observation, background information on the river basin where the hydropower project is located shall be collected, including the hydrographic net of the main stream and its tributaries, multipurpose development, and historical water temperature observation records.

3.0.3 For the prototype observation, the data on project characteristics and project operation, as well as available water temperature data, shall be collected.

3.0.4 For the prototype observation, the topographic maps of the reservoir area and the downstream reaches with a scale no less than 1 : 10 000 shall be collected; and the topographic maps with a scale no less than 1 : 2 000 shall be collected for the water temperature observation sections. The data on large sections in reservoir area and downstream reaches shall comply with the current sector standard NB/T 35094, *Code for Water Temperature Calculation of Hydropower Projects*.

3.0.5 For the prototype observation, the characteristic parameters shall be collected from representative weather stations in the project-affected area, including water temperature, humidity, sunshine hours, wind speed, wind direction, cloud cover, and evaporation.

3.0.6 For the prototype observation, the characteristic parameters shall be collected from representative hydrologic stations in the project-affected area, including flow rate, water level, and water temperature.

3.0.7 When there exist environmentally sensitive objects with special requirements for water temperature, basic data shall be collected, including protected objects, protection levels, function, range, and the requirements for water temperature.

3.0.8 The analysis of basic data for prototype observation of water temperature for hydropower projects shall comply with NB/T 35094, *Code for Water Temperature Calculation of Hydropower Projects*.

4 Technical Requirements for Observation

4.1 General Requirements

4.1.1 Technical requirements for prototype observation of water temperature for a hydropower project shall be coordinated with the existing or other planned water temperature observation of the project. The water temperature observation of reservoir inflow and outflow and the vertical water temperature observation in front of the dam shall be coordinated and consistent.

4.1.2 For the prototype observation of water temperature, the control and representative points shall be selected. The observation sections, lines, and points shall be so arranged as to reflect the spatial distribution of water temperatures in the reservoir and downstream reaches.

4.1.3 The prototype observation of water temperature for hydropower projects shall be carried out for a full hydrologic year. The observation period and frequency shall reflect the variation of water temperature with time in the reservoir and downstream reaches.

4.2 Water Temperature Observation of Reservoir Inflow

4.2.1 For reservoir inflow, observation sections shall be set at the end of backwater in the main stream and key tributaries.

4.2.2 For reservoir inflow, the observation points shall be set 0.5 m below the water surface or the ice cover in the main stream. Where the depth of water is less than 0.5 m, the observation points shall be set at half depth of water.

4.2.3 For reservoir inflow, the observation shall be carried out at 08:00 every day during the observation period. When the daily temperature difference is relatively large, water temperature shall be observed at 08:00, 14:00, and 20:00, respectively.

4.2.4 When real-time observation is adopted for water temperature observation of reservoir inflow, the data recording interval of temperature transducers shall be 1 h or an integral multiple of 1 h. When manual periodic observation is adopted, the temperature sensing and reading of thermometers shall comply with the national standard GB 13195, *Water Quality-Determination of Water Temperature-Thermometer or Reversing Thermometer Method*.

4.3 Vertical Water Temperature Observation of Reservoir

4.3.1 The observation sections for vertical water temperature shall be set in front of the dam and in the reservoir. The observation sections for vertical water temperature in front of the dam shall be set close to the upstream face of the dam. The observation sections for vertical water temperature in the reservoir shall be determined according to the reservoir shape, backwater length, tributary inflow, distribution of environmentally sensitive objects, etc.

4.3.2 At least one vertical water temperature observation line shall be set in the reservoir at the thalweg, and additional observation lines shall be arranged depending on the transverse distribution of water temperature. Water depth sensors shall be set on the observation lines for concurrent water depth observation.

4.3.3 The observation points on the vertical water temperature observation lines shall meet the following requirements:

1 Water temperature observation points shall be set according to the water temperature stratification in the reservoir, and shall be added where the vertical water temperature changes obviously. The water temperature observation points shall so arranged as to reflect the change of vertical water temperature gradient in the reservoir.

2 Water temperature observation points in epilimnion shall be set 0.5 m below the water surface or the ice cover. For the reservoir in a cold region, observation points in epilimnion on the vertical water temperature observation lines shall be added properly according to the ice cover thickness in the frozen period.

3 Water temperature observation points in thermocline should be spaced at 2 m apart along the water depth, and when the temperature difference between adjacent points exceeds 0.40 °C, additional point(s) shall be set in between.

4 Water temperature observation points in hypolimnion should be spaced at 5.0 m to 10.0 m apart along the water depth, and when the temperature difference between adjacent observation points exceeds 0.40 °C, additional point(s) shall be set in between.

5 Water temperature observation points near the reservoir bottom shall be set at 0.1 m to 0.5 m above the mud surface.

4.3.4 Vertical water temperature observation of reservoir shall be implemented during the water temperature rise and drop periods. Water temperature shall be measured at 08:00 every day during the observation period. When the water temperature hysteresis effect is significant in the reservoir, vertical water temperature observation shall also be carried out in both low- and high-temperature periods.

4.3.5 When manual periodic observation is adopted for vertical water temperature observation, the observation interval shall be 2 h, 4 h, 12 h, 24 h, or an integral multiple of 24 h. For the manual observation instrument, the reading of a water temperature sensor at the same observation point shall not be taken until it has stabilized for at least 10 s.

4.3.6 When a thermistor chain is used to observe the vertical water temperature of the reservoir in real time, the floating or fixed thermistor chain shall be adopted depending on the factors such as terrain conditions, hydrological conditions, installation timing, and safety performance. The temperature measurement sensors on the thermistor chain shall be so spaced as to reflect the vertical water temperature gradient.

4.3.7 When there exists icing or temperature inversion distribution in the reservoir, or water temperature observation is required during flood or ecological water discharge, vertical water temperature observation of the reservoir shall meet the following requirements:

1. When there is icing in the reservoir, water temperature observation shall be conducted during the freezing and melting periods, and the ice thickness shall also be measured.

2. When there is temperature inversion distribution in the reservoir, water temperature observation shall be conducted during the water temperature mixing.

3. When water temperature observation is required for the during flood process, water temperature observation shall be conducted during the flood formation, development and receding.

4. Water temperature observation shall be conducted during ecological discharge if required.

4.4 Water Temperature Observation for Selective Water Withdrawal Effectiveness

4.4.1 To check the effectiveness of selective water withdrawal facility, observation sections should be set immediately upstream of the intake and downstream of the tailrace, respectively. For the selective water withdrawal using a temperature control curtain, observation sections shall be set upstream and downstream of the curtain near the vertical centerline, respectively, provided that safety is guaranteed.

4.4.2 To check the effectiveness of selective water withdrawal facility, one vertical water temperature observation line shall be arranged upstream the centerline of the selective water withdrawal facility. When the distribution of the water flow field near the selective water withdrawal facility is complex and the water temperature distribution effect is required to be observed, additional vertical water temperature observation lines shall be provided, and the flow field observation shall also be conducted.

4.4.3 To check the effectiveness of selective water withdrawal facility, the observation points on vertical water temperature observation lines shall be set considering the water withdrawal elevation range. The vertical water temperature shall be observed with emphasis on the top of the selective water withdrawal facility and near the centerline of the water intake. Observation points for tailwater shall be arranged in the mainstream of the tailrace.

4.4.4 The selective water withdrawal effect observation shall last at least one complete operation period of the selective water withdrawal facility, and the water temperature upstream of the selective water withdrawal facility shall be observed at 08:00, 14:00 and 20:00 every day as a minimum during the observation period. When the water temperature upstream of the selective water withdrawal facility is observed in real time using a thermistor chain, the observation shall comply with Article 4.3.6 of this code.

4.4.5 The tailwater temperature observation period and frequency shall be consistent with the water temperature observation upstream of selective water withdrawal facility.

4.5 Water Temperature Observation of Reservoir Outflow and Downstream Reaches

4.5.1 For reservoir outflow temperature observation, observation sections shall be set in the main flowing zones at the outlets of different release structures. When it is required to observe the flood discharge temperature, an observation section shall be set at the control section in the downstream reach where the flood water is mixed.

4.5.2 For water temperature observation of downstream reaches, control sections shall be arranged along the downstream reaches, including the representative sections at existing observation sites, upstream and downstream of key tributary confluences, environmentally sensitive objects, etc. When there is any reservoir in the water temperature influence range downstream of the hydropower project, observation sections shall be arranged at the head and outlet of the downstream reservoir.

4.5.3 When there is water demand for crops irrigation, observation sections shall be arranged along the irrigation channel.

4.5.4 For the water-reduced river reach of a diversion-type or hybrid hydropower station, observation sections shall be arranged at appropriate locations in the water-reduced river reach.

4.5.5 Surface water temperature in the downstream reaches shall be observed, and the technical requirements for the observation shall comply with Article 4.2.2 to Article 4.2.4 of this code.

5 Observation Instruments and Equipment

5.1 General Requirements

5.1.1 Observation instruments and equipment shall be reliable, durable, cost effective, practical, and safe, and should be as advanced as possible to facilitate the automatic collection, transmission, storage, display and sharing of observation data.

5.1.2 The observation instruments and equipment shall be debugged after embedment and installation, and be subjected to a thorough check and calibration before observation, and regular inspection and maintenance during the observation period.

5.1.3 The observation instruments and equipment shall be so arranged as to facilitate installation and maintenance on the premise that they satisfy the technical requirements.

5.2 Configuration of Instruments and Equipment

5.2.1 The observation instruments and equipment shall pass the quality certification, and the measuring range, precision and performance indicators shall meet the corresponding technical requirements. The measuring range of temperature sensor shall be $-5.00\ °C$ to $35.00\ °C$, the resolution shall not be inferior to $0.01\ °C$, the allowable error is $\pm 0.05\ °C$, and its reliability in MTBF shall not be less than 40000 h. The resolution of water depth sensor shall not be inferior to 1.0 cm, the allowable error is ± 3.0 cm when the water level variation is 10.0 m or less, otherwise the maximum allowable error shall be 0.3 % of the measuring range, and its reliability in MTBF shall not be less than 8000 h.

5.2.2 The observation instruments and equipment working underwater, and their components, cables and joints shall have desirable water tightness, water pressure resistance and impact resistance, and shall be able to withstand at least 1.5 times the working water pressure.

5.2.3 The data storage capacity of observation instruments and equipment shall be determined according to the observation period and frequency, and shall meet the storage demand for one year or more.

5.2.4 The signals and interfaces between temperature sensors, displaying and recording instruments, data transmission equipment and other automatic equipment shall be compatible with each other.

5.2.5 The observation instruments and equipment shall conform to the corresponding standards of public network if data is transmitted via postal or digital mobile communication channels, or shall conform to relevant standards for wireless communication circuit design if data is transmitted via ultra-short wave.

5.3 Management of Instruments and Equipment

5.3.1 The observation instruments and equipment shall be calibrated before use, and shall be maintained regularly in use.

5.3.2 The safety, reliability, durability, practicability and automation performance of the observation instruments and equipment shall comply with their respective standards.

5.3.3 The associated communication equipment, computer, power supply, etc. shall comply with their respective standards.

5.3.4 The carrier and mechanical equipment for on-water transportation, operation and safety of the observation instruments and equipment shall comply with their respective standards.

6 Observation Data

6.1 Observation Data Recording

6.1.1 The observation records shall be complete. The records shall include the observation conditions, locations, start and end time, water temperature, water depth, water level, and air temperature. The record form for prototype observation of water temperature for hydropower projects shall comply with Appendix B of this code.

6.1.2 The debugging, calibration, inspection and result comparison of observation instruments and equipment shall be recorded.

6.2 Observation Data Compilation

6.2.1 The observation data shall be compiled timely. The causes of anomalies shall be identified, corresponding measures shall be taken for correction, or re-measurement shall be conducted when possible. When the cause is not clear, the situation shall be described truthfully; nevertheless, the data shall not be altered or removed arbitrarily.

6.2.2 The observation data compilation shall meet the following requirements:

1 Compile the original observation data, observation images and their text descriptions.

2 Check the reasonableness of original observation data and text descriptions.

3 Bind the original observation data and collected references in volumes and make copies.

4 Digitalize and back up the observation data.

6.3 Result Analysis and Data Management

6.3.1 The observation results shall be analyzed for data consistency, reasonableness and completeness according to reservoir characteristics.

6.3.2 The analysis results of observation data should be presented in the form of drawings, tables, time history curves, relation curves, etc.

6.3.3 When conditions permit, the comparative analysis of observation results and calculation results of water temperature shall be conducted.

6.3.4 The observation report shall contain the main body, and attached drawings and tables. For the phased observation, a report shall be prepared for each phase according to the observation scheme, and a summary report shall be prepared after all observation work is finished.

6.3.5 The documents shall be filed in accordance with the national regulations on science and technology archives, including the observation scheme, observation data and analysis results, observation reports, observation contract, terms of reference, and appraisal or acceptance opinion.

Appendix A Contents of Report on Prototype Observation of Water Temperature for Hydropower Projects

1 General

1.1 Background

1.2 Purpose of Observation

1.3 Observation Scope

1.4 Principle of Work

1.5 Process of Work

2 Project Overview

2.1 Overview of River Basin and Environmental Conditions

2.2 Overview of Project

2.3 Overview of Environmentally Sensitive Objects

3 Analysis of Historical Water Temperature Observation Data

4 Observation Scheme

4.1 Observation Range and Period

4.2 Technical Requirements

4.3 Observation Instruments and Equipment

5 Observation Data Compilation

5.1 Basic Data Compilation

5.1.1 Operation Data

5.1.2 Hydrologic Data

5.1.3 Meteorological Data

5.2 Compilation of Reservoir Water Temperature Observation Data

5.2.1 Water Temperature Observation Data of Reservoir Inflow

5.2.2 Vertical Water Temperature Data of Reservoir

5.2.3 Water Temperature Observation Data of Selective Water Withdrawal Effectiveness

5.2.4 Water Temperature Observation Data of Reservoir Outflow

5.3 Compilation of Water Temperature Observation Data of Downstream Reaches

6 Analysis of Observation Results

6.1 Water Temperature Change of Reservoir Inflow

6.2 Reservoir Thermal Structure

6.3 Effectiveness of Selective Withdrawal Facility

6.4 Water Temperature Change of Reservoir Outflow

6.5 Water Temperature Change of Downstream Reaches

7 Conclusions and Recommendations

7.1 Conclusions

7.2 Recommendations

Attached Tables:

1 Statistical Table of Water Temperature Observation Results of Reservoir Inflow

2 Statistical Table of Reservoir Thermal Structure Observation Results

3 Statistical Table of Water Temperature Observation Results of Selective Water Withdrawal Effectiveness

4 Statistical Table of Water Temperature Observation Results of Reservoir Outflow

5 Statistical Table of Water Temperature Observation Results of Downstream Reaches

Attached Drawings:

1 Layout of Prototype Water Temperature Observation Points

2 Water Temperature-Time Curve of Reservoir Inflow

3 Reservoir Thermal Structure

4 Water Temperature-Time Curve of Downstream Reaches

Appendix B Record Form for Prototype Observation of Water Temperature for Hydropower Projects

Table B Record form for prototype observation of water temperature for hydropower projects

Observation location: _____ Section: _____ Observation line: _____

Start date: _____ End date: _____

Description: _____

Sampling time		Water temperature at observation point (°C)	Water depth at observation point (m)	Flow velocity at observation point (m/s)	Flow rate at observation section (m³/s)	Water level at observation section (m)	Air temperature at observation section (°C)	Observation conditions	Remarks
Month	Day	Hour							
...									

Observed by: _____ Date: _____ Recorded by: _____ Date: _____

Explanation of Wording in This Code

1. Words used for different degrees of strictness are explained as follows in order to mark the differences in executing the requirements in this code.

    1) Words denoting a very strict or mandatory requirement:

       "Must" is used for affirmation; "must not" for negation.

    2) Words denoting a strict requirement under normal condition:

       "Shall" is used for affirmation; "shall not" for negation.

    3) Words denoting a permission of a slight choice or in an indication of the most suitable choice when conditions permit:

       "Should" is used for affirmation; "should not" for negation.

    4) "May" is used to express the option available, sometimes with the conditional permit.

2. "Shall meet the requirements of…" or "shall comply with…" is used in this code to indicate that it is necessary to comply with the requirements stipulated in other relative standards and codes.

List of Quoted Standards

GB 13195, *Water Quality-Determination of Water Temperature-Thermometer or Reversing Thermometer Method*

NB/T 35094, *Code for Water Temperature Calculation of Hydropower Projects*